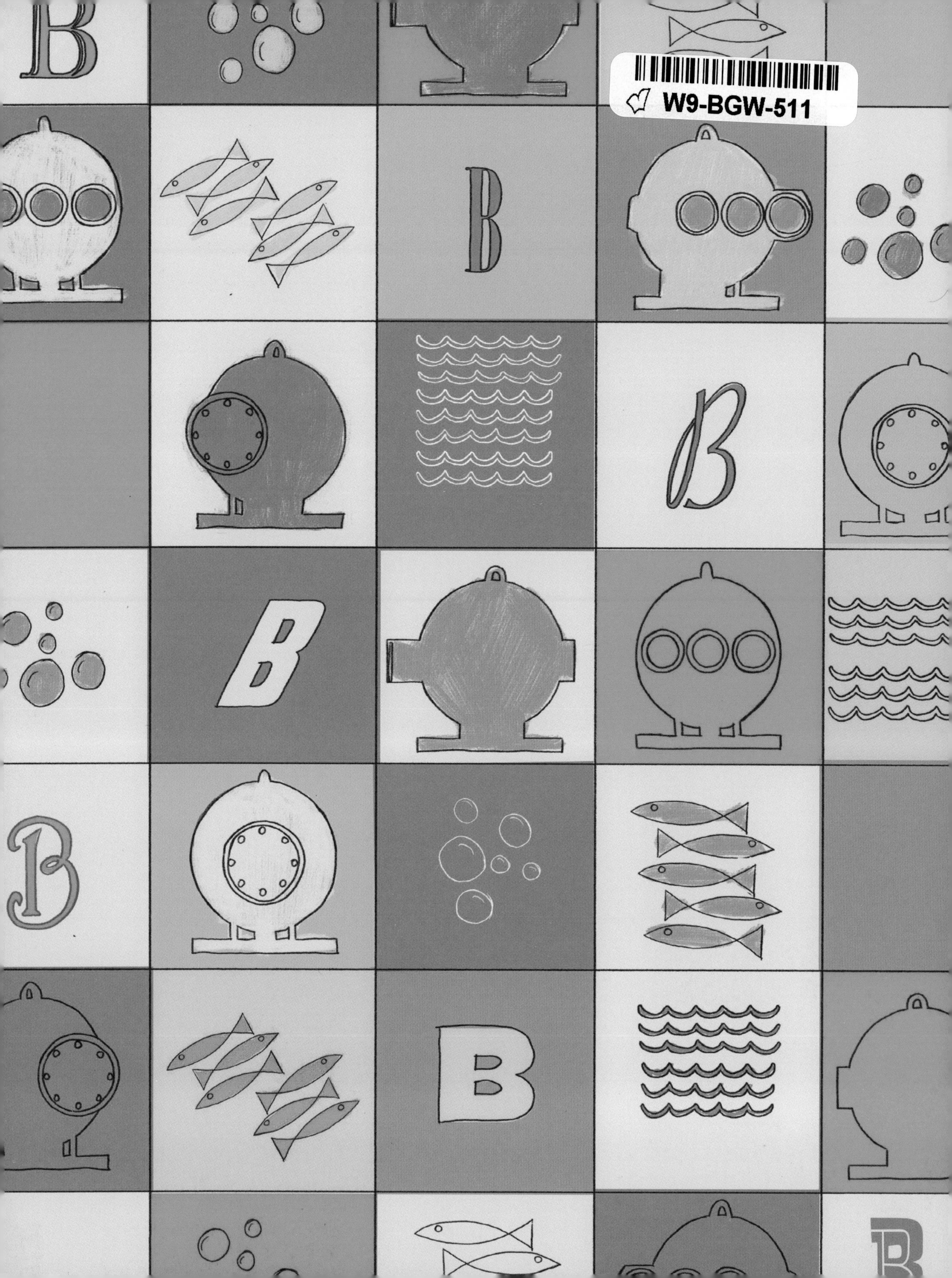

An Unhinged History Book

THE BATHYSPHERE BOYS

Written by Ted Enik
Illustrated by G.F. Newland

MONSTER MAT
TAXI
Coming Soon
TWEEN GOBLIN
NOW PLAYING
HORRORS THAT HIDE
IN A HOLE IN THE SEA
NY 30
2F 59-64

"... the only other place comparable to these marvelous nether regions, must surely be naked space itself, out far beyond atmosphere, between the stars, where sunlight has no grip upon the dust and rubbish of planetary air, where the blackness of space, the shining planets, comets, suns, and stars must really be closely akin to the world of life as it appears to the eyes of an awed human being, in the open ocean, one half mile down."
- William Beebe

"... rather like an enormous inflated and slightly cockeyed bull frog." - Otis Barton

THE BATHYSPHERE BOYS

A Word about the Rhyming

Here and there in *The Bathysphere Boys*, the rhyme words that fall at the end of a line (end rhymes) are close, but slightly "off." This is intentional, like you'll find in the lyrics of many pop and rap songs. It's called *slant, near,* even *lazy* rhyming.

Other paired lines in the book use a technique called "docking"—when I teach my Dr. Seuss class, I call it syllable shifting or swapping. Here's how it works: in a normal pair of rhyming lines (a *couplet*), if one or two syllables are added to the end of a line, they're "docked" from the beginning of the next, but the syllable count remains consistent throughout.

If you're having a little trouble, try reading a few *couplets* out loud. Let the lines' "gallop" take you along. Edgar Allan Poe uses this gallop a lot. Poets call it an *anapest*, which is a kind of accented metrical "foot," or small rhythmic element. An anapest goes bah-dah-BUM, like the words incomplete or seventeen/incom**plete**, seven**teen**.

Here's a line of *anapests* at work in Poe's "Annabel Lee":

"For the **moon** never **beams** without **bring**ing me **dreams** of the **beaut**iful **Ann**abel **Lee**."
Bah-dah-BUM, bah-dah-BUM, bah-dah-BUM, bah-dah-BUM, bah-dah-BUM, bah-dah-BUM, bah-dah-BUM.
You can find many *anapests* in "The Night before Christmas," too.

Here's a few examples from *The Bathysphere Boys*:

After thinking it over, the great man agreed.
And he saw in an instant that Barton succeeded

In solving the shape issue, brilliantly. Clearly,
To better a cylinder, go with a sphere.

After **think**ing it **o**ver, the **great** man a**greed**.
And he **saw** in an **in**stant that **Bart**on suc**ceed**ed

In **sol**ving the **shape** issue, **brill**iantly. **Clear**ly,
To **bet**ter a **cyl**inder, **go** with a **sphere**.

We begin with a prologue, like in an old movie.
A black-and-white flashback all blurry and goofy.

It starts in the dark, "Look! A pinprick of light!"
Zooming in, it's a ship on the high seas at night.

Seems a bomb woke a beast on the floor of the ocean
(. . . it's just a toy dinosaur filmed in slow motion . . .).

Then "Doctor Important" appears from below --
He's your typical science guy; glasses and goatee --

About to be lowered down into the drink.
So the doc gets the dunk and commences to sink.

And we follow him (fearless sardine in a can)
As he's lowered below to a depth any woman

Or man really shouldn't go visit too often --
Especially crammed in a coffee-can coffin.

Cut back to on deck. "Look! The cable's gone slack!"
Yep, the nuclear monster, it snarfed down a snack.
Burp.

A man in a creature suit, two hours later,
Falls--CRASH!--to the ground. Say "Goodbye," giant gator.

What's fun is the history's pretty exact there.
You take out the monster, the rest of it's fact:

* Dashing gentleman, scientist, cowboy-explorer--check.

* Wants to observe deep-sea fauna and flora--check.

* Travels below in a small metal sphere . . .

* And survives! He's an oceanographical miracle man.

Oops, my bad! This deep-sea V.I.P., he
Of course had a name: meet the famous Will Beebe.

And late-1920's New York, the Big Apple,
Is WHEN on a timeline, and WHERE on the map.

NEW YORK

Helmet
Submarine
Diving Suit
Me
Man in a Can

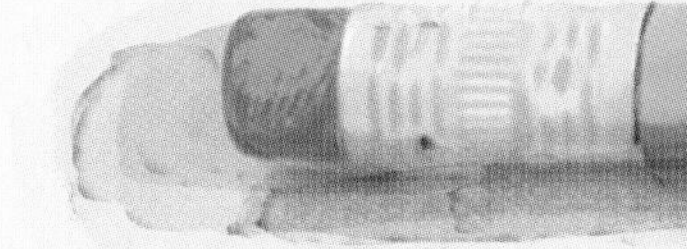

Will, he dreamed about diving, to get his feet wet
At the end of a cable, live marionette-like.

And other than that (it's the '20s we're talking),
The ways to go undersea walking and gawking were few.

* Now a helmet went one-hundred feet,
but beyond that the pressure would pancake you neatly.

* A sub could quadruple that measurement, roughly,
but having no windows made fish-spotting tough.

* A full diving suit added a third-again more,
but because it was clunky and stiff, one who wore it
became pretty much an aquarium statue --
you stand there and hope something cool would swim at you.

And that was it. Either it's those or it's "nope":
Will's a man in a can, or a dope on a rope.

That's why Beebe got busy, filled napkins with sketches
Of all sorts of new submarines that--for stretches

Of time--would allow him to stay under water
And be the world's best-ever deep-sea reporter.

A pro at publicity, Will, he would mention
His diving intention and would-be invention

At lectures, and luncheons, and chance interviews.
And because he was famous, his plans made the news.

Nearly instantly after, Will's desk overflowed
With an unwelcome number of silly proposals

And doodles that, though often wildly inspired,
Were glanced at and straight to the trashcan retired.

Elsewhere in New York, over eggs over easy,
Another man leaped to his feet feeling queasy.

He groaned, Otis Barton did (shoving his dish in),
While rereading, wide eyed, the morning edition

Of the *New York Times* where he happened to notice
A piece about Beebe. "He beat me!" groaned Otis.

A graduate student (Advanced Engineering),
He too had his heart set on--PLOP!--disappearing

Beneath the sea's surface and breaking all records
For deepness, discovering critters and wrecks,

Aboard some sort of special-built diving machine
That explorers--No! Make that THE WORLD's never seen.

And what's more, Barton saw in a half of a *snap*
That Will's favorite design, it belonged on the scrap-heap.

It wouldn't survive with that soda-can shape,
Under deepwater pressure, things crumple like paper!

So Barton persuaded a mutual friend
To go visit with Beebe, convince him to spend

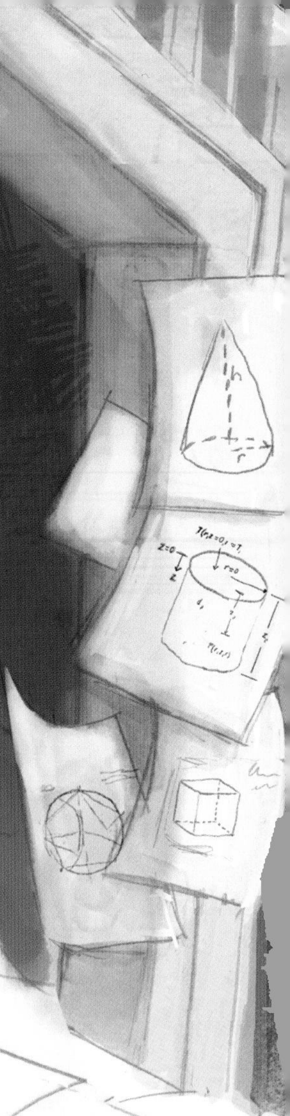

Twenty minutes with Otis, to hear his idea.
Or risk being one-upped by the young engineer.

After thinking it over, the great man agreed.
And he saw in an instant that Barton succeeded

In solving the shape issue, brilliantly. Clearly,
To better a cylinder, go with a sphere.

And on top of that, Beebe discovered the Bartons
Of Boston were rich! So not only a smart one

Was Otis, he also--convenient and neatly --
Was willing to pay for the project completely.

He'd happily do so with one small request:
If he builds The Ball, he joins the dive as a guest.

In the thrill of the moment, their future in gear,
Beebe christened their baby the first "Bathysphere!"

My Design
Ring!
Ring!

From New York
to Nonsuch Island

Bon Voyage!

Now as everyone knows:
* Oil and water won't mix.
* Pouring milk into lemon juice makes something "icky."
If "opposites" ever you needed defining,
Our Otis and Will--as examples--were shining.

'Cause Beebe thought Barton was lightweight and lazy;
Disliked him for being celebrity crazy.

And Barton thought Beebe was stuffy, a bore:
He resented Will's fame, and how Nerd-Girls adored him.

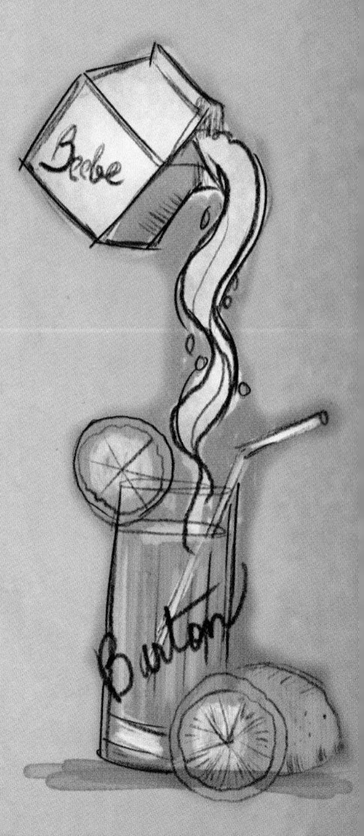

BROOKLYN ELECTRICAL
SALES RECEIPT
Jan 1930

But all that aside, Otis--not one for stalling--
Flipped open his wallet and got The Ball rolling.

With metal machinists, and iron refiners,
With glass blowers, rope makers, lighting designers.

When finished, The Sphere carried two people--barely.
And looked like a coconut minus the hair.

“She'll be snug,” Barton joked, “there'll be no room for crackers.”
In total, it cost the man 12,000 smackers!

In June, on the 6th, 1930 the year,
Was the first-ever dive of the first Bathysphere!

The Ready

And they knew in an instant, did Beebe and Otis,
They must stay in touch with the folks on the boat.

For it's not hard to fathom, the farther they sank
For a telephone line they were terribly thankful.

Will, nose to a porthole, began his narration
(A mixture of "Wow!" and profound observation):

A detailed tally, an "any-and-all" list
Of all that he witnessed, to Gloria Hollister.

Who, up on deck, on their barge, took it down
Word for word. That's including the clowning around.

Then at 300 feet, the sea came sort of flowing
In under the hatchway, but still they kept going.

Next, something electrical shorted out, throwing
Off sparks (and a sharp stink), and STILL they kept going.

At 800 feet, Beebe jumped when a strong
Feeling hit him that stuff was about to go wrong.

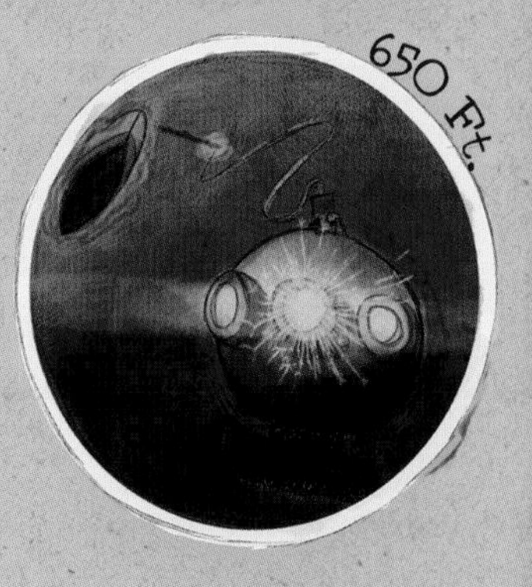

So he ordered "Now! No delay!" that they should scuttle
The dive. Will, he usually trusted his gut.

Yeah, but "details, details . . ." Nobody died.
Neither fellow got waterlogged, flattened, or fried.

CELEBRATION!

Their cargo-barge chugged back to shore
With its sirens up full, and its fog horns 'a-roaring!

When summertime '31 came to a close,
Both our Bathysphere heroes were deep-diving pros.

'Cause their big metal yo-yo had so often seen
What's below (and come back!) it became a routine.

So for Gloria H., their First Mate on the phone?
On her birthday they gave her a dive of her own!

And for dozens of years she alone held the crown
For the farthest a female had ever been down.

And then SPLAT!!! The Depression (the "great" one) descended:
Their funding dried up, going under--BANG!!!--ended.

The whole of the nation stayed broke as can be,
So no diving was done in 1933.

But The Bathysphere never got junked. No, it rather
Stayed famous and likewise continued to gather up

Bravos, and kudos, and more than its share
Of admirers at the Chicago's World's Fair.

Where it went on exhibit (Will stood by and eyed it,
To shoo away kids who kept crawling inside).

Ah, but getting attention from tourists is one thing . . .
What Will needed most was a way to get funding.

Now back then, some magazines, sensing a story,
Would sponsor explorers to share in their glory.

So Will swore to write something thrilling and fabulous.
Nat' Geographic said, "We'll pay the tab!" For his

One
Last
Big
Dive (and it wouldn't come cheap).

They agreed--if he went down a HALF MILE DEEP!!!

It was August 15th, 1930-and-4
When The 'Sphere, like a satellite, left the shore's orbit.

Successfully so! Till a mountainous wave
Rocked the barge, and in turn caused The Ball to behave

Like a pendulum, which convinced Barton to vomit --
Dramatically. Or, quoting Will, "Like a comet."

Despite the smell (Mr. Barf, welcome aboard!),
Both explorers agreed that the "show" should go forward.

'Cause Otis, he wanted to capture on camera
The sea's weirdest creatures; their features, their glamour.

And turn what he shot into some sort of movie.
A deep-sea adventure film, that's what he'd do.

Even though his companion, he couldn't care less
About anything other than watery blackness.

Will gasped about shapes bobbing just out of range
Of The Bathysphere's headlight--amazingly strange.

And he gushed over light flashes, showers of sparks
Given off by the luminous life in the darkness.

At 2,000 feet a world record was broken!
Praised every-which-where such successes were spoken of.

Also all newspapers and magazines
Spread the word that The Ball Boys had beat submarines . . .

To the basement-est,
cellar-est,
level-the-lowest

That humankind, up to now, ever did goest!

Then all of a suddenly, CLUNK!--"What was that?!?!"
Otis overheard folks on the barge sounding rattled

And worried. The next moment--YANK!--he and Beebe
Were being hauled up 'cause a cable tore free!

That was close! Nearly killed after so many dives,
That you might think the sea felt they owed her their lives . . .

There's no argument; Otis and Will from the start,
Weren't buddies. They merely put up with the partnership.

What broke the camel's back once and for all?
It was Barton releasing his cheesy, and awful,

And really embarrassing mess of a flick.
Beebe said the few seconds he saw made him sick.

It contained:

* Fuzzy footage from some of their dives, mixed together with
* Sea critters (some of them live . . .),
* A tame shark fight with big-helmet girls in bikinis,
* And scenes from a filmstrip about submarine-ies

That made mighty warships look like bathtub toys.

And that . . .
was the end . . .
of the Bathysphere Boys!

It was also the end of The 'Sphere's work in science.
They stuck it in storage like some old appliance.

Until 1939's New York World's Fair,
Where The Ball got a hall of its own and an airing out --

Once again people could come and admire it --
Nautical hero come out of retirement.

When World War II hit, our Navy's attention
Was turned to new deepwater weapon invention;

To make the best boat-sinking thingamajig,
And The 'Sphere was recruited to play guinea pig.
BOOM!

We bandaged the planet, we ended The War,
Then the battle-scarred Bathysphere had the good fortune

To find a good home at the New York Aquarium,
In Coney Island; the nonstop World's Fair.

The amazing thing is, nearly 40 years later,
Is that's where they wasted that nuclear gator!

Time-jumping ahead, it's 1994,
"Let's spruce up The Aquarium!" orders a foreman.

The Ball gets the boot (like that out-of-date toaster),
And joins all the junk they dump under their Coaster.

BRONX
ZOO
New York
Zoological Society
MANHATTAN
QUEENS
New York World's Fair 1939
BROOKLYN
(Birthplace of Will Beebe)
New York Aquarium
Coney Island
START
END
(Nuclear Gator's Grave)
R.I.P
(Deep-Water Blasting)
USS

But dough, and the tide, and the sun always rises,
And fate, for The 'Sphere, had a few more surprises.

It's rescued! Then polished, repainted, repaired.
Even gets to grow old in a brand-new aquarium!

Like a prize pearl--those are one of a kind,
Or the egg of a dinosaur--fabulous find,
Or a big Christmas ornament--megasize fun,
Or a meteor resting--its skydiving done . . .

Overlooking the lunatic Coney-Baloney,
The Bathysphere sits now.
A queen
on her throne.

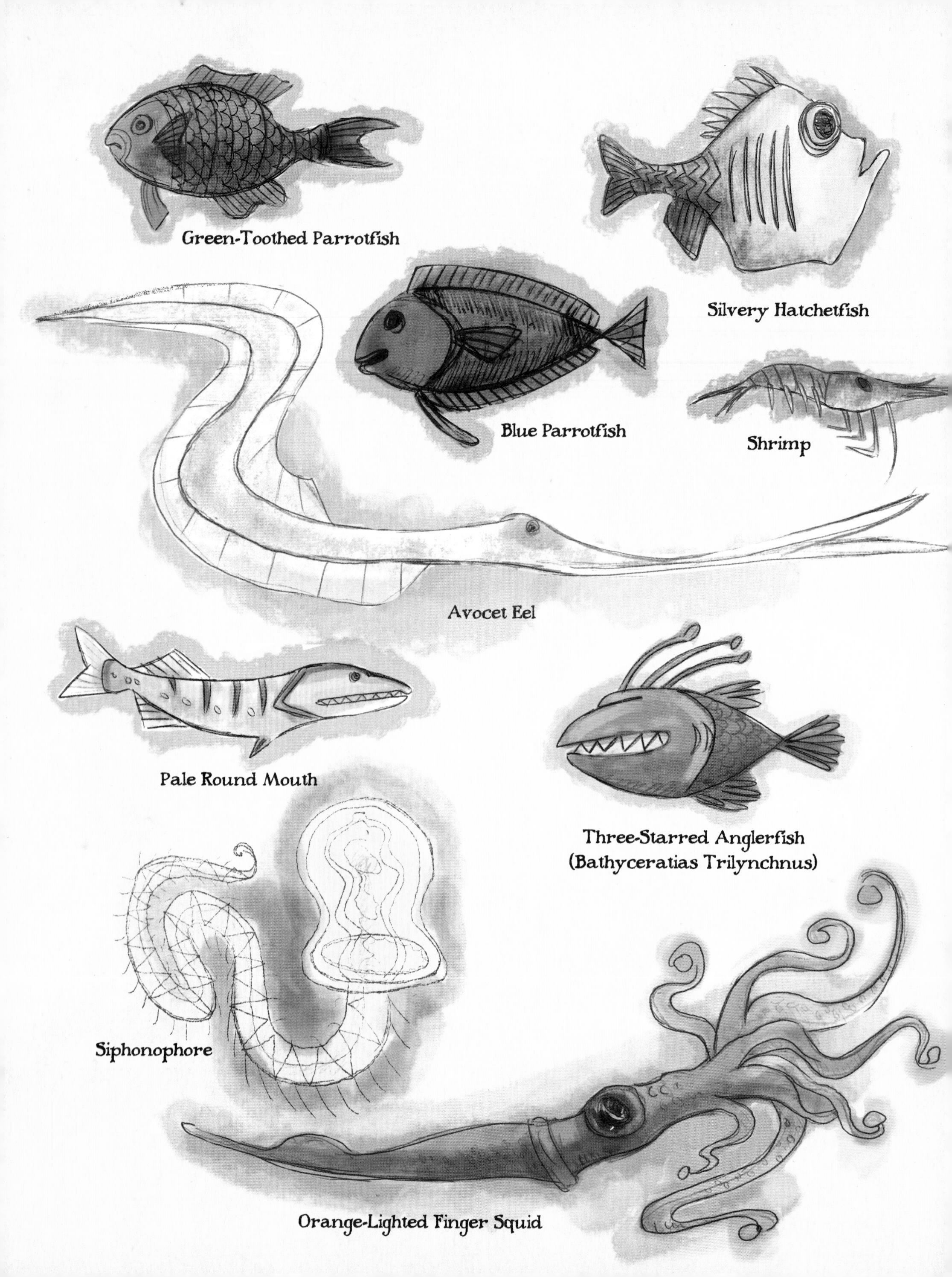
Green-Toothed Parrotfish
Silvery Hatchetfish
Blue Parrotfish
Shrimp
Avocet Eel
Pale Round Mouth
Three-Starred Anglerfish
(Bathyceratias Trilynchnus)
Siphonophore
Orange-Lighted Finger Squid

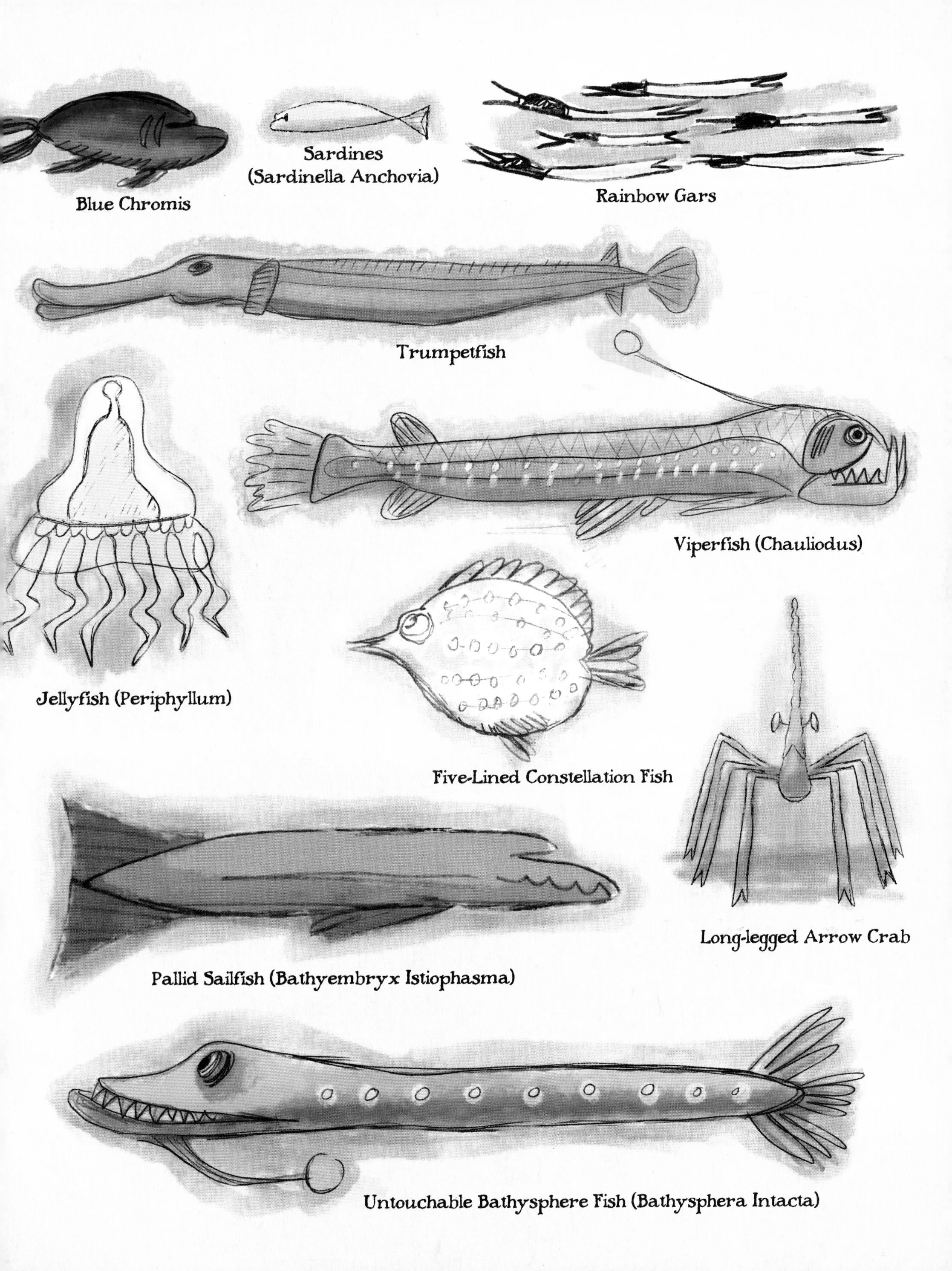
Blue Chromis
Sardines
(Sardinella Anchovia)
Rainbow Gars
Trumpetfish
Jellyfish (Periphyllum)
Viperfish (Chauliodus)
Five-Lined Constellation Fish
Long-legged Arrow Crab
Pallid Sailfish (Bathyembryx Istiophasma)
Untouchable Bathysphere Fish (Bathysphera Intacta)

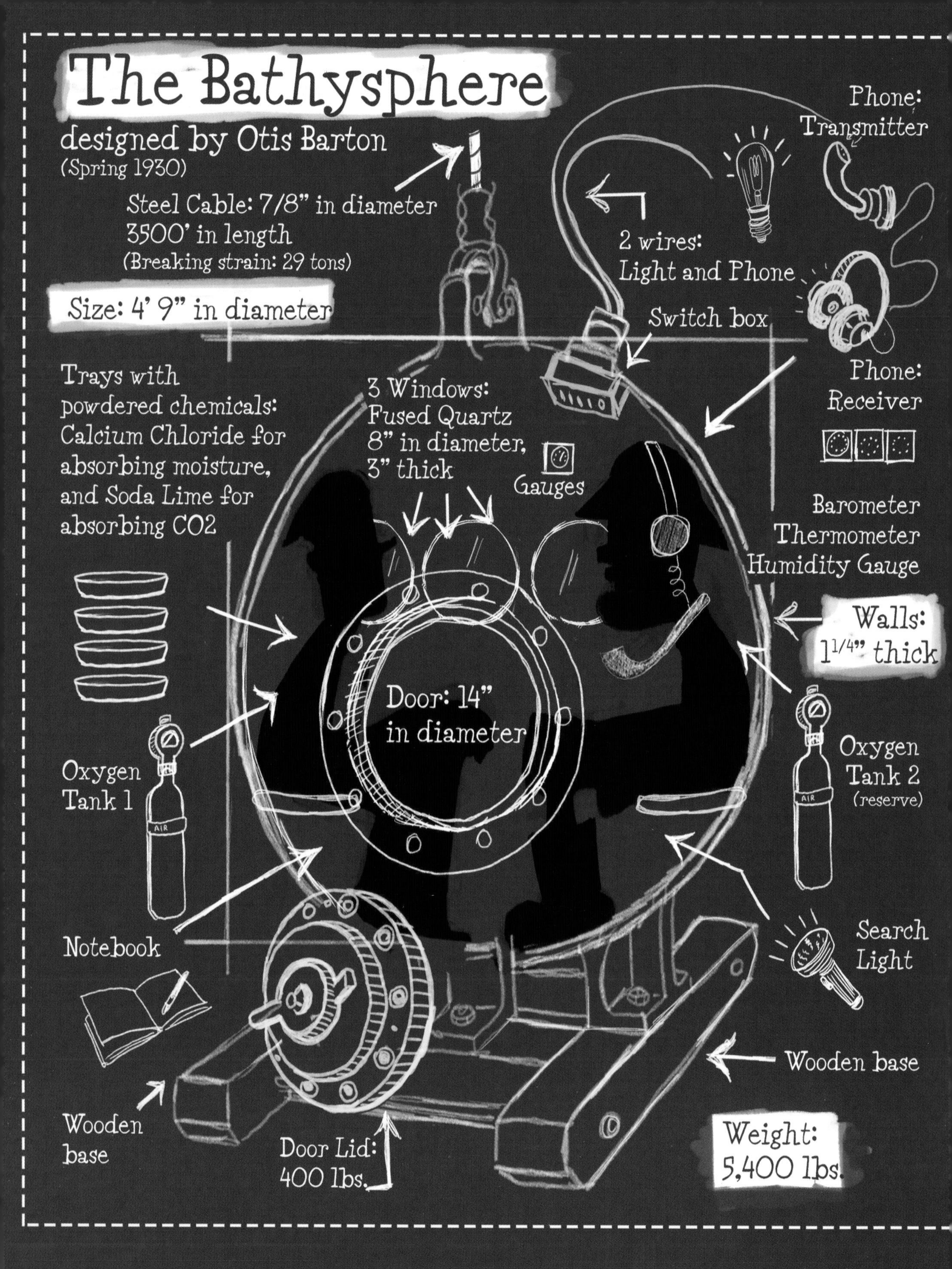

The Bathysphere
designed by Otis Barton
(Spring 1930)
Steel Cable: 7/8" in diameter
3500' in length
(Breaking strain: 29 tons)
Size: 4' 9" in diameter
Phone: Transmitter
2 wires: Light and Phone
Switch box
Phone: Receiver
Trays with powdered chemicals: Calcium Chloride for absorbing moisture, and Soda Lime for absorbing CO2
3 Windows: Fused Quartz 8" in diameter, 3" thick
Gauges
Barometer
Thermometer
Humidity Gauge
Walls: 1 1/4" thick
Door: 14" in diameter
Oxygen Tank 1
AIR
Oxygen Tank 2 (reserve)
AIR
Notebook
Search Light
Wooden base
Wooden base
Door Lid: 400 lbs.
Weight: 5,400 lbs.

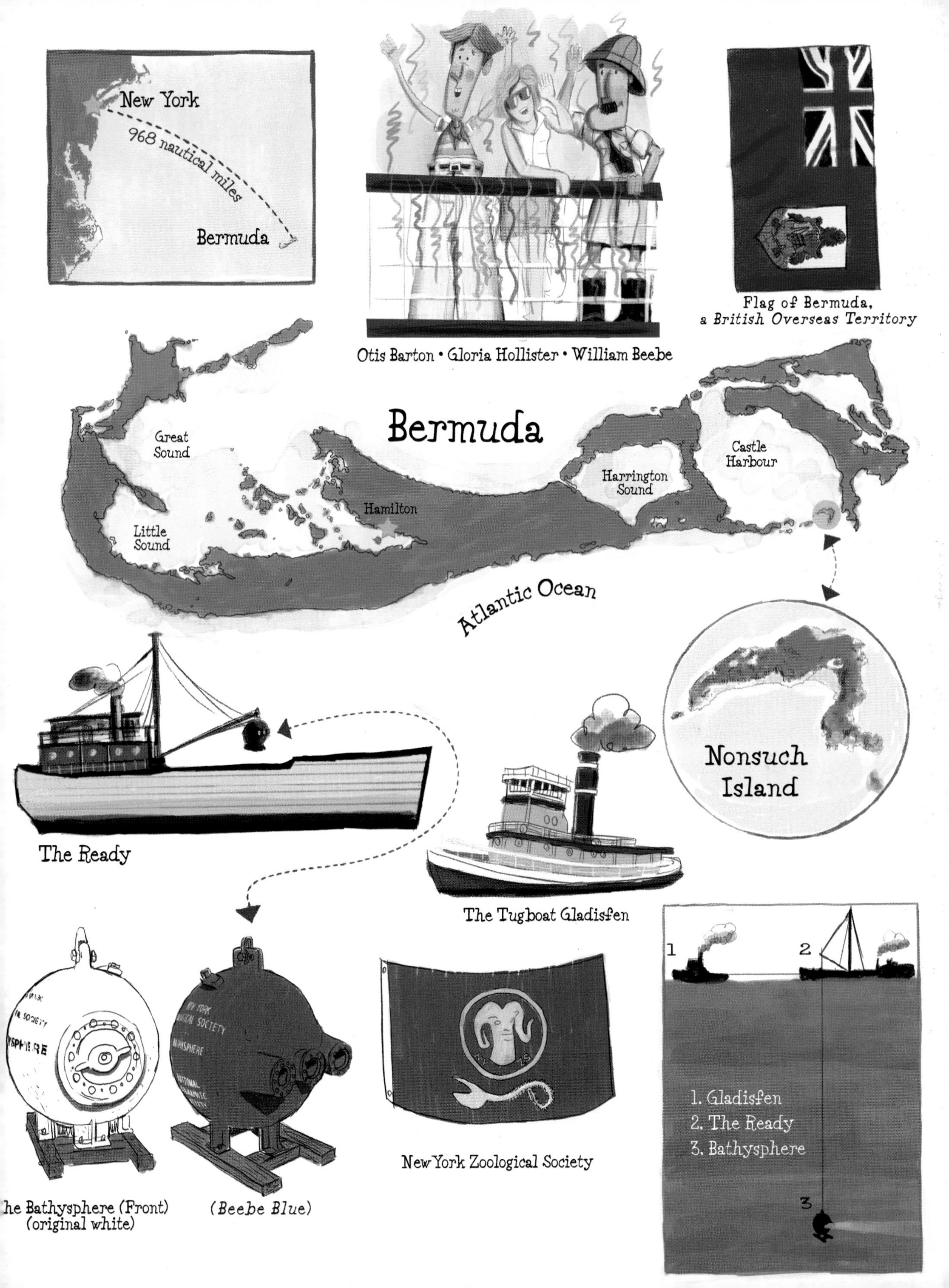
New York
968 nautical miles
Bermuda
Otis Barton • Gloria Hollister • William Beebe
Flag of Bermuda,
a British Overseas Territory
Bermuda
Great Sound
Little Sound
Hamilton
Harrington Sound
Castle Harbour
Atlantic Ocean
Nonsuch Island
The Ready
The Tugboat Gladisfen
1
2
3
1. Gladisfen
2. The Ready
3. Bathysphere
New York Zoological Society
he Bathysphere (Front)
(original white)
(Beebe Blue)

THE EVOLUTION OF HUMAN DIVING

(a Rough, Unreliable Timeline of basically, the Tubes, Cans, Boxes, and Bells that came before the Bathysphere.) Chart adapted from "Half Mile Down" by William Beebe (1934), New York: Harcourt, Brace and Company. Thank you, Prof. Beebe!

1
2
3
4
5

❶ **REED BREATHING** Really old school. Could go as deep as a human is tall, plus the length of the reed.

❷ **TRAPPED-AIR BUBBLE DOME** Inspired by the Water Spider's air-filled, silk "Bell." Could go a little deeper than a human is tall. More freedom of movement.

❸ **LEATHER DIVING HELMET** Inventor Unknown. Early 16th century. Could go about 4 ft. deep above the average human's height.

❹ **LORINI'S DIVING "TRUMPET"** 17th century. Could go about 25 ft. deep.

❺ **EDMOND HALLEY'S "DIVING BELL"** Could go about 50 ft. deep. Early 18th century.

❻ **JOHN LETHBRIDGE'S DIVING "PAPOOSE"** Could go about 60 ft. deep. Early 18th century.

❼ **KARL HEINRICH KLINGERT'S early "modern" DIVING SUIT.** Could go as deep as a riverbed. Late 18th century.

❽ **THE DIVING SUIT** Several Inventors are credited:
* August Siebe's "Heavy Footer." Early 19th century.
* Alphonse and Theodore Carmagnolle's "Portholes." Mid-19th century.
* Charles Macintosh's "Waterproof Diving | Outfit." Late 19th century.
All could go about 200 ft. deep, more or less.

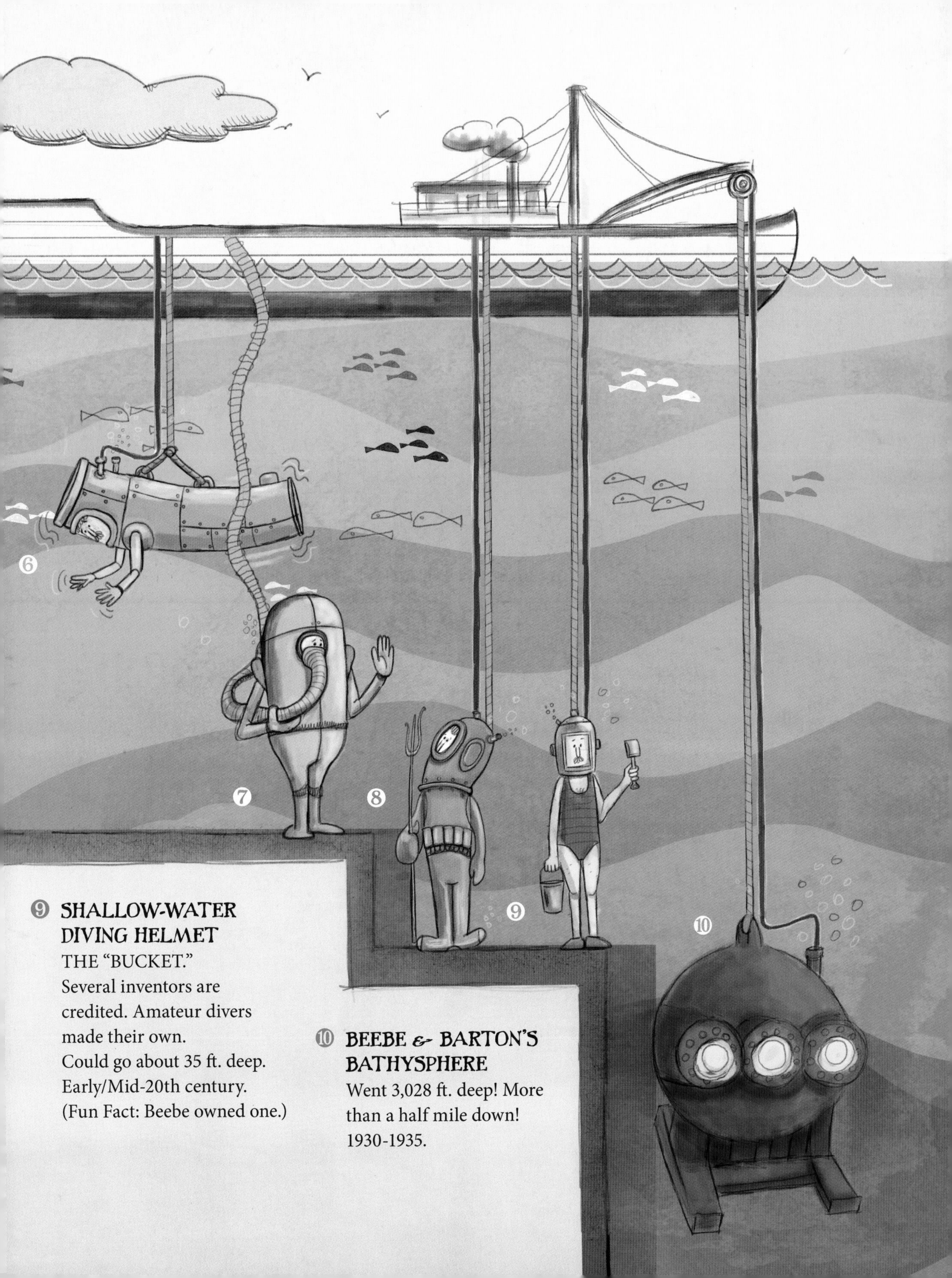

⑨ SHALLOW-WATER DIVING HELMET

THE "BUCKET."
Several inventors are credited. Amateur divers made their own.
Could go about 35 ft. deep.
Early/Mid-20th century.
(Fun Fact: Beebe owned one.)

⑩ BEEBE & BARTON'S BATHYSPHERE

Went 3,028 ft. deep! More than a half mile down!
1930-1935.

Gloria Hollister

Explorer, Scientist, Conservationist.

Miss Hollister, as she preferred to be called, became a science celebrity in the 1930s, a rare thing in the mostly male world of exploration and discovery.

Born in New York City, as a girl Gloria spent summers at the Hollister country home upstate. There her father sparked what became a lifelong curiosity for the natural world by letting her care for fancy poultry and assist with dissections of dead farm animals.

In 1924, Gloria graduated from Connecticut College for Women, where she was class president, twice, and excelled as a high jumper, discus thrower, and a member of the All-America girls' hockey team. Soon after, the athlete earned a master's in zoology at Columbia University and became research assistant to the famed oncologist Alexis Carrel.

Grown tired of the lab and longing to work outdoors, in 1928 Hollister applied for a job with William Beebe at the New York Zoological Society. Beebe needed an accomplished naturalist, an expert at dissection, to join his Department of Tropical Research (DTR). Despite fellow explorers advising him against a "girl scientist," Beebe immediately recognized Hollister's skills and hired her.

On a DTR expedition to Bermuda, Beebe, Barton, and Hollister explored the deep sea in the Bathysphere. On her 30th birthday she went down 410 feet, breaking the women's record for the deepest dive. In '34, she nearly tripled that by reaching 1,208 feet — both unique opportunities to explore the unknown. Hollister also invented a way to make fish specimens transparent, allowing their skeletons to be studied more easily.

Soon after, Hollister headed her own DTR expedition to British Guiana's Kaieteur Falls, five times taller than Niagara. Skimming the dense jungle in a small plane, she discovered over 40 new waterfalls, most known only to native peoples.

In later life, Hollister continued to collaborate with Beebe and other scientists at the Zoological Society. She became a member of the Society of Women Geographers, earning their Outstanding Achievement Award, helped establish America's first blood donor organization in Brooklyn, and became a lecturer-ambassador of the American Red Cross.

Throughout a long and brilliant career, Gloria Hollister humbly called herself a "detective seeking nature's secrets."

Other Schiffer Books by Ted Enik and G.F. Newland:
Sticks 'n' Stones 'n' Dinosaur Bones: Being a Whimsical "Take" on a (pre) Historical Event, Ted Enik, Illustrated by G.F. Newland, ISBN: 978-0-7643-5394-9

Library of Congress Control Number: 2019936468

Edited by Kim Grandizio

Type set in Gorey/Minion

ISBN: 978-0-7643-5793-0
Printed in China

Co-published by Pixel Mouse House & Schiffer Publishing, Ltd.
4880 Lower Valley Road | Atglen, PA 19310
Phone: (610) 593-1777; Fax: (610) 593-2002
E-mail: Info@schifferbooks.com | Web: www.schifferbooks.com

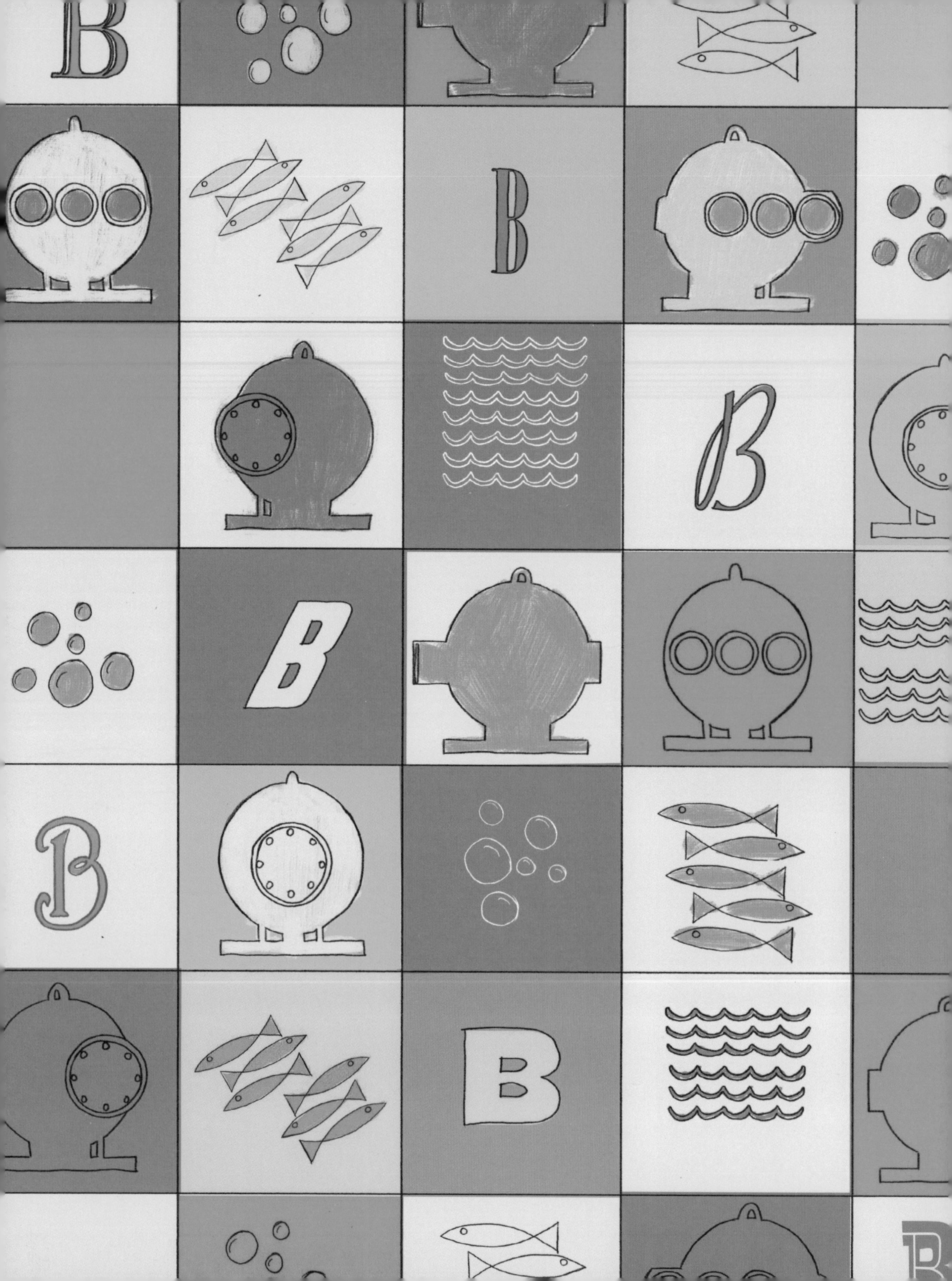